WHALES

WHALES

AN *illustrated* CELEBRATION

KELSEY OSEID

TEN SPEED PRESS

California | New York

CONTENTS

INTRODUCTION

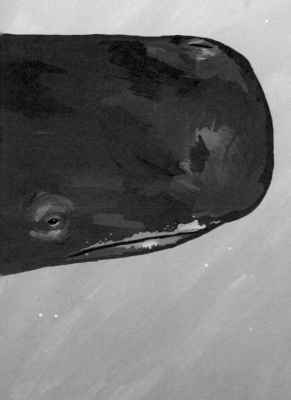

Whales, dolphins, and porpoises have fascinated humans for millennia. Cetaceans are, like us, mammals. But they exist in a completely different world. We live on land, they live in water. We travel north, east, west, south—but cetaceans add "up" and "down." Still, in many ways we're not so different. We breathe air, we nurse our young, and we have complex family lives—just as cetaceans do.

This book is an illustrated exploration of cetaceans' evolutionary history, taxonomy, behaviors, and more. It covers how whales, dolphins, and porpoises evolved from early mammals and came to dominate as the giants of the modern sea. You'll get to know cetaceans' unique mammalian bodies, along with their incredibly intelligent minds. We'll examine the vast diversity of cetacean species swimming the earth's oceans and rivers today, and offer a glimpse into the inner workings of their complex lives (though parts of their lives are mysterious and largely unknown to us humans). And we'll dive in to some of the ways we as humans have connected with and related to our cetacean cousins—including, sadly, the ways we've exploited them for our own gain.

Renowned marine biologist Dr. Sylvia Earle once said, "Even if you never have the chance to see or touch the ocean, the ocean touches you, with every breath you take, every drop of water you drink, every bite you consume." Our lives as land mammals are fundamentally tied to the sea, and at this critical time in the world's history—when humans are taking more from the earth than it can give us—we owe it to the sea to appreciate and try to understand it. Learning about the world's fascinating cetaceans—those intelligent, complex, sometimes caring, sometimes ferocious, deep-diving, traveling mammals of the sea—allows us to examine our interconnection with the world's water, and the life within it.

VOCABULARY

The taxonomic group Cetacea is divided into two major categories: the toothed whales (odontocetes) and the baleen whales (mysticetes). We tend to think of "whales" as a category separate from "dolphins," but dolphins and porpoises are actually kinds of toothed whales.

There are a number of terms and phrases that are helpful to know when studying and learning about whales, dolphins, and porpoises.

BULL
An adult male cetacean.

CALF
A baby cetacean.

COW
An adult female cetacean.

POD
A group of cetaceans (sometimes also called a "school" or a "herd").

THE GREAT WHALES

A term used to refer to the biggest cetaceans. The baleen whales referred to in this category vary, but typically include the bowhead whale, the right whales, the gray whale, the blue whale, the fin whale, Bryde's whale, the minke whales, and the humpback whale. The largest toothed whales—the sperm whale and the three giant beaked whales—may also be included in this informal category.

WHALES

Often refers to the larger cetaceans of any family, or can refer specifically to species whose common names include "whale," like the pilot whales, rorqual whales, and so on.

DOLPHINS

An informal category containing cetacean species not generally considered whales or porpoises.

PORPOISES

Cetaceans belonging to the family *Phocoenidae* are smaller than most other cetaceans. While most toothed whales and dolphins have cone-shaped teeth, porpoises have spade-shaped teeth.

ANATOMY

Cetaceans' highly specialized anatomy has evolved to perfectly suit their needs in their aquatic habitats. Some parts of their bodies (such as fins) may be familiar, since they are shared by many other marine species, but other features (like blowholes and melons) are unique to cetaceans.

BLOWHOLE
The respiratory opening on the top of a cetacean's head, analogous to the nostril of other mammals.

MELON
The bulging forehead of some toothed whales, which focuses sounds and aids in echolocation.

BEAK
A cetacean's forward-projecting jaws. Also referred to as the snout.

DORSAL FIN
The raised fin on the back of most, but not all, species of cetacean.

FLIPPERS
The two paddle-shaped front limbs of a cetacean. Also called the pectoral fins.

FLUKES
Cetacean tails are made up of two horizontally flattened lobes, each called a fluke.

CETACEANS HAVE STREAMLINED BODIES LIKE SHARKS AND OTHER FISH, BUT THEIR TAILS MOVE UP AND DOWN INSTEAD OF SIDE TO SIDE

BALEEN WHALES HAVE TWO BLOWHOLES—
TOOTHED WHALES HAVE ONE

CALLOSITY
Raised, roughened skin on the head of a right whale.

ROSTRUM
The upper jaw (sometimes refers to both upper and lower jaws).

SPLASHGUARD
A raised area in front of the blowhole of some whales that prevents water from entering the blowhole.

TAIL STOCK
The muscular structure between a cetacean's dorsal fin and flukes; also referred to as the "peduncle" or "caudal peduncle."

BLUBBER
A thick layer of fat under the skin in cetaceans and other marine mammals.

BALEEN
The fringed plates in certain whales' mouths that allow them to filter feed, straining their food from seawater. Baleen is also referred to as "whalebone," even though it is not made of bone but of keratin, the same substance that makes up our fingernails.

MANDIBLE
The lower jaw.

5

SURFACING BEHAVIORS

The easiest cetacean behaviors for humans to observe are the ones performed at the surface of the water. Some of these behaviors have clear functional purposes—logging, for example, is believed to be a way for certain species to rest while continuing to breathe at the surface. Other surface behaviors are multipurpose—while cetaceans may tail-slap or breach to communicate, or porpoise to build up speed as they travel, they also may perform these behaviors purely for fun.

LOGGING
A form of rest in which the cetacean floats motionless at the surface.

FLIPPER-SLAPPING
Slapping the flippers against the surface of the water. Also called pectoral-slapping.

PORPOISING
Leaping above the surface of the water while swimming, often to gain speed.

SPY HOPPING
Popping the head just above the surface of the water, often to look around.

WAKE-RIDING
Swimming in and along the wake behind a ship.

BOW-RIDING
Swimming in and along the waves coming off the bow of a ship.

BREACHING
When a cetacean leaps partially or completely out of the water (a leap that exposes less of the body is sometimes called a "lunge" or a "surge").

TAIL-SLAPPING
Slapping the flukes against the surface of the water. Also called lobtailing.

WHALE BLOWS

Because they are air-breathing mammals, cetaceans must periodically rise to the surface of the water and take in fresh air. When large whales do this, it causes a "blow"—a visible spray of water vapor. Different species have very distinct blows, and the shape, size, and angle of the blow can help identify whales at sea. And because whales blow as they surface and right before they dive, a blow can be a useful tip-off for whale watchers that a whale is about to be visible and possibly to show its flukes.

RAINBLOW!
If the light catches a blow just right, it can cause a rainbow.

The gray whale has a bushy blow that looks V-shaped when viewed head-on.

The bowhead also has a distinctive V-shaped blow, since its two blowholes are relatively far apart.

The humpback whale has a wider, bushier blow.

The blue whale's blow is a straight column that can be over thirty feet tall.

The sperm whale's blowhole is on the left side of its head, so its blow angles forward and to the left.

FLUKING

"Fluking" is the term for when a whale raises its tail flukes above the surface of the water right before a dive. Some whales lift their flukes entirely out of the water, others only partly do, and some don't raise their flukes out of the water at all. Different species have distinct fluke shapes, too; both the shape of the flukes and the way they are presented above the water can help whale watchers identify different species at sea.

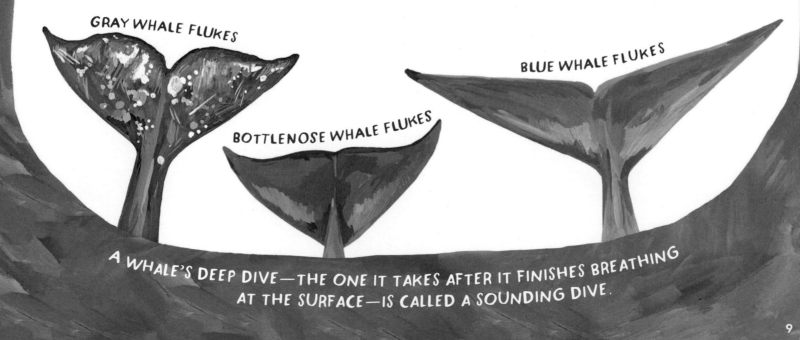

GRAY WHALE FLUKES

BOTTLENOSE WHALE FLUKES

BLUE WHALE FLUKES

A WHALE'S DEEP DIVE—THE ONE IT TAKES AFTER IT FINISHES BREATHING AT THE SURFACE—IS CALLED A SOUNDING DIVE.

BEACH-RUBBING

When cetaceans swim close to shore or shallows to rub their bodies against sandy or pebbly surfaces it's called beach-rubbing. This can be a social activity for some cetaceans. Beluga whales in the Arctic, for example, have been documented traveling hundreds of miles to gather in freshwater estuaries with particularly well-suited pebbly bottoms, where they exfoliate and clean themselves by beach-rubbing.

STRANDING

A whale, dolphin, or porpoise that ends up on land is "stranded" or "beached." In some parts of the world, including the U.S., it is illegal to touch a stranded whale, dolphin, or porpoise—first aid must be given by authorized professionals. In other places, whale strandings may be treated as sacred sites: for the Maori people of New Zealand, for example, whales are considered sacred, and treaties between the Maori and New Zealand's government allow the traditional harvesting of body parts and other resources from stranded dead whales.

Toothed whales are much more likely to become stranded than baleen whales, and some species are much more likely to strand than others. Among the most likely to strand are pilot whales, which are highly social and tend to strand in groups; strandings of several hundreds of pilot whales at a time are not uncommon.

The causes of individual and mass strandings are not well understood, but may have something to do with weather, disease, cetaceans' use of magnetic fields in navigation, or any number of other factors.

DISTRIBUTION

There are five major oceans that span the world: the Arctic, Atlantic, Indian, Pacific, and Southern. Cetaceans are found in nearly every part of every ocean, from the deep open ocean to coastlines, and from the poles to the equator. Here are a few terms used to describe the distribution of different cetacean species—that is, the places they are typically found in the oceans.

CIRCUMPOLAR
Distributed around the northern or southern poles. The bowhead whale, found in the Arctic, is an example of a circumpolar species.

COASTAL
Distributed in and around coastal areas. Burmeister's porpoise, found exclusively along the coast of South America, is an example of a coastal species.

COSMOPOLITAN
Distributed more or less all over the world. Killer whales are an example of a cosmopolitan species.

PELAGIC
Found in the waters of the open oceans, far from coasts. Cuvier's beaked whale is an example of a species that spends its time in pelagic areas.

PACIFIC OCEAN

ARCTIC OCEAN

Some species have incredibly small
ranges, like the vaquita, which lives
only in a small area in the northern Gulf of California.
Other species have incredibly wide ranges—humpback
whales, for instance, can traverse thousands of miles
between their hunting and breeding grounds, which
they travel between many
times throughout the
course of their lifetimes.

ATLANTIC OCEAN

PACIFIC OCEAN

INDIAN OCEAN

SOUTHERN OCEAN

EVOLUTION

We're currently living in the **Cenozoic era**, which spans the sixty-five million years since the end of the Cretaceous period and the extinction of the dinosaurs. When reptilian dinosaurs dominated the planet, small mammals did exist, but they hadn't yet evolved to large sizes. Most were small and shrew-like, and they weren't top predators. The dawn of the Cenozoic era and the extinction of the dinosaurs opened up a space for mammals to grow larger and dominate the landscape. We often think of the early "age of mammals" as a time of woolly mammoths and saber-toothed cats, but it wasn't just the age of mammals on land—it was also the age of **mammals in the sea**, when the very first whales began to evolve.

Cetaceans belong to the kingdom Animalia, in the phylum Chordata, in the class Mammalia. As mammals, they are warm-blooded, give birth to live young, and nurse their offspring with milk. They are one of the most widely distributed groups of mammals on Earth, with species living as far north as the Arctic Circle and as far south as the waters of the Antarctic, and in all of the rest of the world's oceans, as well as many smaller seas, bays, rivers, and tributaries.

This sprawling, diverse group of sea mammals all descended from an early mammalian ancestor. At one time, that common ancestor was believed to be the wolf-like *Sinonyx*. Today, cetaceans are believed to have descended from early artiodactyls—creatures from the order Artiodactyla, which today includes modern even-toed ungulates (aka hoofed animals) like giraffes, deer, and pigs. (Even-toed means their weight is borne equally by the third and fourth toes.) The earliest cetaceans appeared more than fifty million years ago.

WADI AL-HITAN

Wadi Al-Hitan (Arabic for "Valley of the Whales") is the name of an important paleontological site in Egypt that is home to an incredible diversity of hundreds of archaeoceti fossils. It's a desert now, but it was once the site of the prehistoric Tethys Sea, home to proto-whales like **Basilosaurus** and **Dorudon**.

KING LIZARD

"Basilosaurus," meaning "king lizard," sounds like the name of a dinosaur, not a mammal. When scientists first discovered the remains of Basilosaurus, they understandably mistook them for those of a reptile. Even though we now know the species to be a mammalian whale relative, the scientific name for this genus remains the same.

EARLY WHALE RELATIVES

Though we haven't identified any direct ancestors of whales, we have found evidence of many species that were their early relatives. The earliest known cetacean relatives still had hind legs (as in *Pakicetus*). As they adapted to live more and more aquatic lives, those hind legs became flippers (as in *Dorudon*). These early relatives of the cetaceans are referred to as the Archaeocetes.

Adaptations to aquatic life included an auditory system adapted for better hearing underwater.

INDOHYUS
A close relative of cetaceans.

PAKICETUS
Evolved around fifty-two million years ago.

Adaptations to aquatic life included legs adapted to wading in water.

Adaptations to aquatic life included a completely internal hearing system, with no external ears.

AMBULOCETUS
Name means "walking whale."

Evolved around forty-eight million years ago.

KUTCHICETUS
Evolved around
forty-six million
years ago.

Adaptations to aquatic life included a stronger tail
that aided in swimming while back limbs shrank.

Adaptations to aquatic life included nostrils that have
moved to the top of the head, becoming a blowhole.

DORUDON
Evolved around forty million years ago.

BASILOSAURUS
Evolved around
forty million years ago.

Adaptations to aquatic life included
back limbs functionally gone; probably
had small flukes to assist in swimming.

OCEAN GIANTS

Why and how did modern-day cetaceans grow to such extreme sizes? The blue whale—the largest modern cetacean—is the largest animal on Earth. It's thought to be larger than any animal ever to have existed on Earth; with a maximum weight of 173 metric tons, it is bigger even than any known land or marine dinosaur. While the other great whales trail behind the blue whale in size, many still rival the sizes of the largest dinosaurs.

To put the massive size of the largest cetaceans in perspective, the largest living land mammal is the African bush elephant (*Loxodonta africana*). The largest noncetacean species swimming the oceans today is the whale shark (*Rhincodon typus*), a filter-feeding cartilaginous fish species that can reach maximum lengths of 41 feet (12.5 meters). The blue whale puts this size to shame, at more than double the maximum length of the whale shark.

BLUE WHALE
Up to 98 feet (30 m) long

SHONISAURUS (EXTINCT)
Up to 49 feet (15 m) long

WHALE SHARK
Up to 41 feet (12.5 m) long

SPERM WHALE
Up to 67 feet (20.5 m) long

MEGALODON (EXTINCT)
Up to 60 feet (18 m) long

GIANT SQUID
Up to 43 feet (13 m) long

GREAT WHITE SHARK
Up to 24 feet (7 m) long

APATOSAURUS (EXTINCT)
Up to 75 feet (23 m) long

AFRICAN BUSH ELEPHANT
Up to 20 feet (6 m) long

OTHER MARINE MAMMALS

Marine animals live in oceans and coastal estuaries. Cetaceans as a group are generally considered marine mammals, though some can and do live in fresh water. And they're not the only mammals to adapt to life in the water—pinnipeds, sirenians, sea otters, and polar bears also live marine lives.

ALL MAMMALS THAT LIVE IN THE WATER EVOLVED FROM MAMMALS THAT LIVED ON LAND

SIRENIANS

Four species, including three species of manatee and the dugong.

The closest living relatives to these rotund herbivores are, surprisingly, elephants and small furry mammals called hyraxes.

SPECIAL MARINE ADAPTATIONS INCLUDE a tapered and rounded body shape to prevent drag while swimming.

SEA OTTERS

Enhydra lutris

The sea otter belongs to the taxonomic family *Mustelidae*, along with other species of otter, badgers, weasels, minks, and more. But unlike any of its *Mustelid* cousins, it is adapted for a fully marine lifestyle.

SPECIAL MARINE ADAPTATIONS INCLUDE the thickest fur of any mammal, to insulate their bodies against the cold water they live in.

POLAR BEARS

Ursus maritimus

Like the sea otter, the polar bear is from a family of nonmarine animals, the *Ursids*. Uniquely among bears, the polar bear is adapted to—and dependent on—a life in and around the sea.

SPECIAL MARINE ADAPTATIONS INCLUDE large, slightly webbed feet to aid with swimming and walking on thin polar ice.

PINNIPEDS

Thirty-three species, including fur seals, harbor seals, elephant seals, sea lions, walruses, and more.

The pinnipeds are relatives of dogs, cats, bears, and other carnivorans. Most pinniped species prefer life in the colder marine waters of the north and south.

SPECIAL MARINE ADAPTATIONS INCLUDE the blubber that helps them keep warm in water and acts as an energy store.

NEW DISCOVERIES AND REMAINING MYSTERIES

We're always learning new information about the taxonomy and evolutionary history of cetaceans. Beaked whales, because of their remote, deep-sea habitats, have been a mystery to scientists for many years. Some species have yet to be conclusively identified in the wild, and many are known only from stranded bodies that have washed up on beaches. New species are being identified all the time, but for now the beaked whales remain largely a mystery.

As we continue to study cetaceans, and technology continues to improve, there's no doubt that further discoveries will yet be made. It's exciting to think there are species swimming the depths of the oceans that have yet to be identified; and there is the possibility of missing links in the evolutionary history of our fellow mammals still undiscovered!

THE MYSTERY OF THE PYGMY RIGHT WHALE

The rarely-seen-at-sea pygmy right whale is another mysterious species. It shares its common name with fellow mysticetes the right whales, but its features are very different, and its evolutionary origins were unknown for most of the time humans have studied it. A 2012 study suggested that the pygmy right whale is actually the last living member of a cetacean family long thought to be extinct: the cetotheres.

ANATOMY THAT REVEALS HISTORY

Cetacean flippers appear perfectly adapted to the life of a marine mammal from the outside, with their smooth, paddle-like shapes. But on the inside, their skeletons reveal their evolutionary history— they still have the individual finger bones that they inherited from their early mammal ancestors.

MODERN CETACEAN RELATIVES

Today, cetaceans are classified as part of the taxonomic order Artiodactyla, the even-toed ungulates. Cetaceans don't have hooves, but they did evolve from the same ancestors as modern artiodactyls, and they retain vestigial leg bones—they're not externally visible in a live whale or dolphin, but you can see them in their skeletons. Hippopotamuses—semiaquatic members of the order Artiodactyla—are believed to be the closest living relatives of cetaceans.

THE **LARGEST OF THE MYSTICETES** IS
THE **BLUE WHALE**, WHICH CAN REACH UP TO
98 FEET (30 M) LONG

THE SMALLEST IS THE PYGMY
RIGHT WHALE, WITH TYPICAL
LENGTHS OF 18 TO 21 FEET
(5.5 TO 6.5 M)

THE **LARGEST OF THE ODONTOCETES** IS THE
SPERM WHALE, WHICH CAN REACH A LENGTH
OF 59 FEET (18 M)

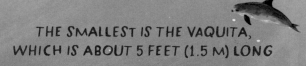

THE SMALLEST IS THE VAQUITA,
WHICH IS ABOUT 5 FEET (1.5 M) LONG

The way species are classified is
always changing, especially as genetic
technology improves. This chapter is not
an authoritative texonomy of cetacean
species, but rather a general overview of
most of the species that are considered
part of the cetacean clade.

CHAPTER TWO

THE SPECIES

Modern cetaceans are divided into two major groups:
Mysticeti, the baleen whales, and **Odontoceti**, the toothed whales.

MYSTICETES

This group includes the right whales, the gray whale, and the rorquals. All mysticetes have baleen, a specially adapted structure that lines their mouths, through which they filter seawater to trap their prey. Many mysticete species have been threatened by industrial-scale hunting by humans, who once used blubber for energy and baleen to make a wide variety of commercial products. Today, thanks to conservation efforts, some species have recovered, but others, like the North Pacific right whale, are still endangered.

ODONTOCETES

Unlike the mysticetes, odontocetes have teeth. They tend to eat larger prey than the mysticetes do, and they have a special adaptation called "echolocation," which allows them to use sound to hunt and communicate. Dolphins and porpoises are odontocetes—in a sense, they are just as much "whales" as their larger relatives. In fact, some members of the odontocete family *Delphinidae* (the oceanic dolphins), are as large as or larger than some of the smaller baleen whales. Like the baleen whales, the toothed whales have faced threats from human hunting.

BALEEN

TOOTH

BALAENIDAE THE RIGHTS AND BOWHEAD

With some of the largest bodies in the animal kingdom, the bowhead takes its name from its bow-shaped skull; the other three species also have dramatically arched upper jaws. All lack dorsal fins. They use their baleen, the longest of any species, for skim feeding. Because of their immense amounts of blubber and their desirable baleen, these whales were major targets of modern industrial whaling, which has left many populations of *Balaenidae* endangered to this day.

BELIEVED TO HAVE THE LONGEST LIFESPAN OF ANY SPECIES OF MAMMAL—CAN LIVE MORE THAN 200 YEARS

BOWHEAD WHALE *Balaena mysticetus*

NAMED FOR ITS CURVED, BOW-SHAPED SKULL

SOUTHERN RIGHT WHALE *Eubalaena australis*

CALLOSITIES FORM
DIFFERENT PATTERNS
ON EACH INDIVIDUAL

200 TO 270 BALEEN
PLATES ON EACH
SIDE OF MOUTH

THICKEST BODY FAT OF ANY
SPECIES—ITS BLUBBER CAN BE
20 INCHES (50 CM) THICK

33

BALAENIDAE THE RIGHTS AND BOWHEAD
CONTINUED

KNOWN FOR THEIR
GIGANTIC HEADS, WHICH
MAKE UP AROUND A THIRD OF
THEIR TOTAL BODY LENGTH

NORTH PACIFIC RIGHT WHALE
Eubalaena japonica

NORTH ATLANTIC RIGHT WHALE *Eubalaena glacialis*

TOTAL BODY LENGTH
OF UP TO 59 FEET (18 M)

AMONG THE MOST ENDANGERED
WHALE SPECIES, WITH ONLY A FEW
HUNDRED INDIVIDUALS REMAINING

PYGMY RIGHT WHALE
Caperea marginata

CETOTHERIIDAE
THE PYGMY RIGHT WHALE

This whale gets its name from some superficial characteristics it shares
with the true right whales, but it was a mystery to scientists for many
years. Recent studies have suggested that it is actually the last living
species of the genus *Cetotherium*, which was previously believed to have
died out during the Pliocene Epoch several million years ago.

BALAENOPTERIDAE THE RORQUALS

The whales of the family *Balaenopteridae* are known as rorqual whales. Rorquals have specially adapted throat pleats that allow the throat to expand to take in vast amounts of water that is then filtered through their baleen, trapping prey in their mouths. The blue whale can take in 20,000 pounds (9,000 kg) of food and water at a time. Many rorqual species suffered exploitation by industrial whaling. Thanks to whaling bans and other conservation efforts, some species, like the blue whale and humpback whale, have largely recovered, but others, like the sei whale, are still endangered.

BLUE WHALE *Balaenoptera musculus*

THE SECOND-LARGEST SPECIES OF WHALE

THE **LARGEST**
ANIMAL ON EARTH

LONG GROOVES IN THE
THROAT ENABLE IT TO
STRETCH TO ENGULF PREY

ITS TONGUE CAN
WEIGH AS MUCH
AN ELEPHANT

FIN WHALE *Balaenoptera physalus*

HAS AN ASYMMETRICALLY COLORED JAW, WHICH IS
LIGHT GRAY ON THE RIGHT AND BLACK ON THE LEFT

BALAENOPTERIDAE THE RORQUALS
CONTINUED

KNOWN FOR ITS SLEEK SHAPE
AND GRACEFUL SWIMMING STYLE

THE SMALLEST
OF THE RORQUALS

PRONOUNCED
"MINK-EE"

COMMON MINKE WHALE
Balaenoptera acutorostrata

FLIPPERS HAVE
DISTINCTIVE
WHITE BANDS

38

SEI WHALE *Balaenoptera borealis*

ANTARCTIC MINKE WHALE
Balaenoptera bonaerensis

SLIGHTLY
BIGGER THAN
THE COMMON
MINKE

39

BALAENOPTERIDAE THE RORQUALS
CONTINUED

PRONOUNCED
"BREW-DUH'S"

BRYDE'S WHALE *Balaenoptera brydei*

CAN BE IDENTIFIED BY THE
THREE DISTINCTIVE RIDGES
ALONG ITS ROSTRUM

ADULTS CAN GROW
TO 50 FEET (15.5 M)
IN LENGTH

OMURA'S WHALE *Balaenoptera omurai*

PREVIOUSLY CONSIDERED A PYGMY FORM OF BRYDE'S WHALE, BUT IS NOW CONSIDERED ITS OWN SPECIES

ONE OF THE SMALLER RORQUAL SPECIES

BALAENOPTERIDAE THE RORQUALS
CONTINUED

While the rest of the rorquals belong to the genus *Balaenoptera*, the humpback belongs to a genus all its own, *Megaptera*. Considered the most energetic and acrobatic of the large whales, it is known for its leaping, breaching, flipper slapping, and tail throws. It also has one of the longest mammal migrations in the world, swimming 10,000 miles (16,000 km) or more on round trips between its calving and feeding grounds. Its distinctive flippers are the largest in relation to its body of any cetacean.

NAMED FOR THE HUMP IN FRONT OF THE DORSAL FIN

HUMPBACK WHALE *Megaptera novaeangliae*

MALES SING THE LONGEST AND MOST COMPLEX SONGS IN THE ANIMAL KINGDOM

FLUKE "FINGERPRINTS"

HUMPBACK WHALES HAVE DISTINCTIVE BLACK-AND-WHITE MOTTLED PATTERNS ON THEIR FLUKES

THESE MARKINGS ARE SO UNIQUE THAT PHOTOS OF THEIR FLUKES ALONE CAN BE ENOUGH TO TRACK AND IDENTIFY INDIVIDUAL HUMPBACK WHALES

43

ESCHRICHTIIDAE THE GRAY WHALE

This family contains only one species: the gray whale. It has the shortest baleen of all the mysticetes, which it uses to filter sediment it sucks up from the sea floor. Their migrations rival the humpback's—they may travel more than 12,000 miles (19,000 km) in their journeys from breeding grounds to feeding grounds. Historically, they were sometimes known as "devilfish" because the mothers are extremely protective of their young, and when whalers came between them, the mothers would defend their calves with wild attacks. Since protections have been put in place and whaling has declined, gray whales have begun to trust people again and are now known for often being friendly and interested in humans.

GRAY WHALE *Eschrichtius robustus*

SINCE IT FEEDS ON ITS RIGHT, ITS BALEEN IS MORE WORN DOWN ON THAT SIDE

INSTEAD OF A DORSAL FIN, GRAY WHALES HAVE A HUMP AND A ROW OF DORSAL "KNUCKLES"

ITS SKIN IS A MOTTLED GRAY, OFTEN DOTTED WITH BARNACLES AND SCARRING

45

DELPHINIDAE THE OCEANIC DOLPHINS

The largest cetacean family, *Delphinidae*, contains some thirty species throughout the world's oceans. Delphinids are generally smaller than most whales, though this family has some larger species, too, like the orca. Most delphinids have a dorsal fin, which can vary in shape across species, from tall and pointed to short and rounded to hook-like. Some have prominent beaks, while others have shorter or indistinct beaks. Many species live in large pods.

SHORT-BEAKED
COMMON DOLPHIN

Delphinus delphis

Delphinus capensis

LONG-BEAKED
COMMON DOLPHIN

A SMOOTH, CONICAL FOREHEAD GIVES THIS SPECIES THE NICKNAME "SLOPEHEAD"

ROUGH-TOOTHED DOLPHIN
Steno bredanensis

PRONOUNCED "TOO-COO-SHEE"

SOME TUCUXI LIVE IN THE FRESH WATER OF THE AMAZON BASIN

PROMINENT HUMP BENEATH THE DORSAL FIN

TUCUXI *Sotalia fluviatilis*

INDO-PACIFIC HUMPBACKED DOLPHIN

Sousa chinensis

COLOR RANGES FROM PINK TO WHITE TO GRAY IN DIFFERENT POPULATIONS AND INDIVIDUALS

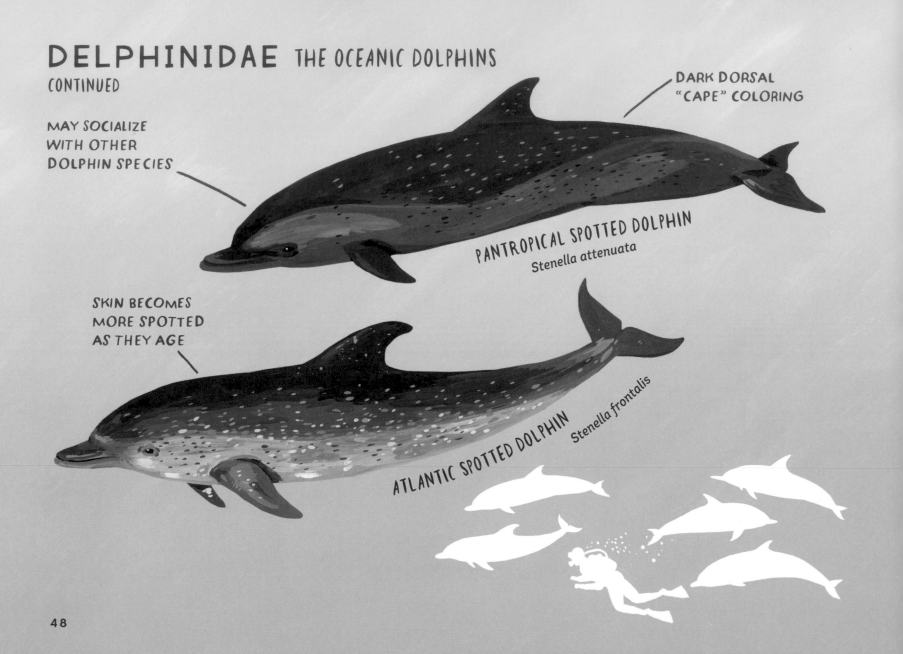

DELPHINIDAE THE OCEANIC DOLPHINS
CONTINUED

DARK DORSAL
"CAPE" COLORING

MAY SOCIALIZE
WITH OTHER
DOLPHIN SPECIES

PANTROPICAL SPOTTED DOLPHIN
Stenella attenuata

SKIN BECOMES
MORE SPOTTED
AS THEY AGE

ATLANTIC SPOTTED DOLPHIN
Stenella frontalis

SPINNER DOLPHIN

LEAPS OUT OF THE WATER AT HIGH SPEEDS AND SPINS BEFORE CRASHING BACK DOWN

Stenella longirostris

CLYMENE DOLPHIN

Stenella clymene

CAN LEAP AND SPIN LIKE A SPINNER DOLPHIN

STRIPED DOLPHIN

Stenella coeruleoalba

DISTINCTIVE STRIPED MARKINGS

DELPHINIDAE THE OCEANIC DOLPHINS
CONTINUED

KNOWN TO SOCIALIZE WITH
OTHER SPECIES OF SEA LIFE

COMMON BOTTLENOSE DOLPHIN

Tursiops truncatus

OFTEN SEEN
LEAPING, WAKE
RIDING, AND
BREACHING

INDO-PACIFIC
BOTTLENOSE DOLPHIN

Tursiops aduncus

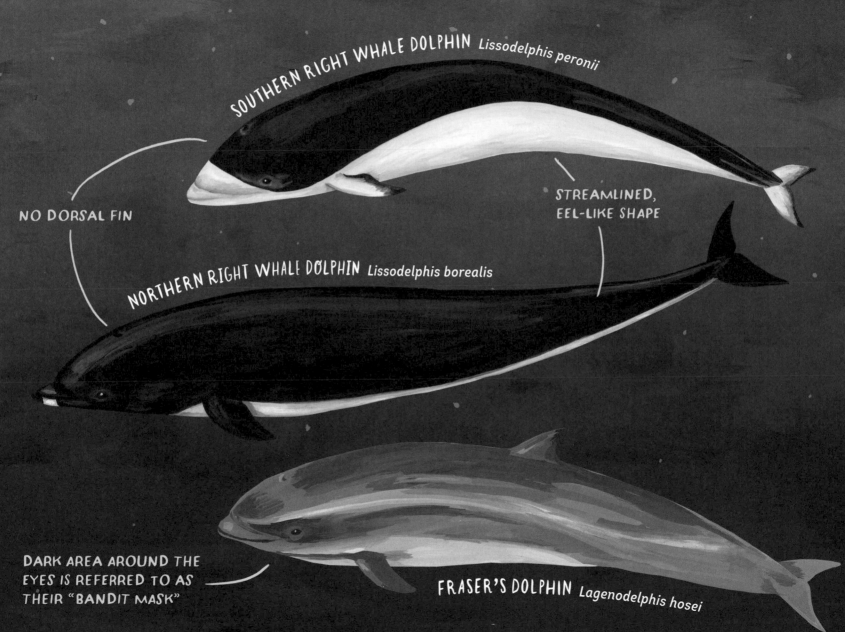

SOUTHERN RIGHT WHALE DOLPHIN *Lissodelphis peronii*

NO DORSAL FIN

STREAMLINED, EEL-LIKE SHAPE

NORTHERN RIGHT WHALE DOLPHIN *Lissodelphis borealis*

DARK AREA AROUND THE EYES IS REFERRED TO AS THEIR "BANDIT MASK"

FRASER'S DOLPHIN *Lagenodelphis hosei*

DELPHINIDAE THE OCEANIC DOLPHINS
CONTINUED

COMMERSON'S DOLPHIN

Cephalorhynchus commersonii

CONICAL HEAD WITH
NO DISTINCT BEAK

AN ACROBATIC
SWIMMER, KNOWN TO
SPIN UNDERWATER AND
BOW- OR WAKE-RIDE

CHILEAN DOLPHIN

Cephalorhynchus eutropia

ALSO CALLED THE
BLACK DOLPHIN

HEAVISIDE'S DOLPHIN

Cephalorhynchus heavisidii

MAY LIVE ALONGSIDE RISSO'S DOLPHINS AND NORTHERN RIGHT WHALE DOLPHINS

HECTOR'S DOLPHIN

Cephalorhynchus hectori

FOUND IN THE COASTAL WATERS OF SOUTHWEST AFRICA

THE SMALLEST SPECIES IN THE OCEANIC DOLPHIN FAMILY

DELPHINIDAE THE OCEANIC DOLPHINS
CONTINUED

A PARTICULARLY ACROBATIC SPECIES—
ONE OF ITS FAVORITE MOVES INVOLVES
LEAPING OUT OF THE WATER AND SPINNING
HEAD-OVER-TAIL BEFORE REENTERING

DUSKY DOLPHIN

Lagenorhynchus obscurus

PACIFIC WHITE-SIDED DOLPHIN

Lagenorhynchus obliquidens

MAY GATHER IN
GROUPS OF UP TO
2,000 INDIVIDUALS

54

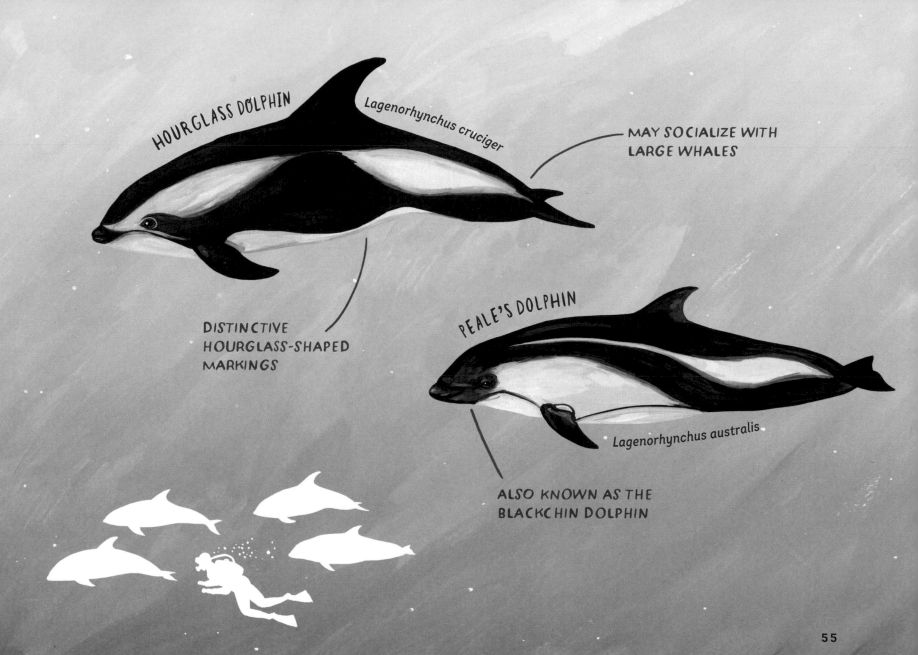

HOURGLASS DOLPHIN

Lagenorhynchus cruciger

MAY SOCIALIZE WITH
LARGE WHALES

DISTINCTIVE
HOURGLASS-SHAPED
MARKINGS

PEALE'S DOLPHIN

Lagenorhynchus australis

ALSO KNOWN AS THE
BLACKCHIN DOLPHIN

DELPHINIDAE THE OCEANIC DOLPHINS

CONTINUED

FALSE KILLER WHALE Pseudorca crassidens

SHORT FLIPPERS LOOK
LIKE THEY'RE BENT

LONG-FINNED PILOT WHALE

Globicephala melas

ALONG WITH THE KILLER
WHALE, THESE DARK-COLORED
DOLPHINS ARE ALSO KNOWN
AS BLACKFISH

THE SPECIES
MOST LIKELY TO
MASS STRAND

Globicephala macrorhynchus

SHORT-FINNED PILOT WHALE

EATS MOSTLY SQUID

THE TWO SMALLEST BLACKFISH SPECIES ARE SOMETIMES MISTAKEN FOR ONE ANOTHER

MELON-HEADED WHALE
Peponocephala electra

PYGMY KILLER WHALE
Feresa attenuata

57

DELPHINIDAE THE OCEANIC DOLPHINS
CONTINUED

HEAVY SCARRING ON SKIN, USUALLY
CAUSED BY OTHER RISSO'S DOLPHINS

RISSO'S DOLPHIN

ALSO KNOWN
AS GRAMPUS

Grampus griseus

IRRAWADDY DOLPHIN

WHITE-BEAKED
DOLPHIN

Orcaella brevirostris

A CLOSE RELATIVE
OF THE KILLER WHALE,
THOUGH THEY LOOK
VERY DIFFERENT

ATLANTIC
WHITE-SIDED   DOLPHIN

Lagenorhynchus albirostris

Lagenorhynchus acutus

DESPITE ITS NAME,
ITS BEAK COLOR
RANGES FROM WHITE
TO DARK GRAY

OFTEN SEEN SPENDING
TIME WITH OTHER DOLPHIN
AND WHALE SPECIES

KILLER WHALE *Orcinus orca*

TALL DORSAL FIN, ESPECIALLY IN MALES AND OLDER WHALES

THE LARGEST SPECIES OF OCEANIC DOLPHIN

ALSO KNOWN AS THE ORCA

BECAUSE OF THEIR TENDENCY TO HUNT AS A PACK, THEY ARE SOMETIMES CALLED "THE WOLVES OF THE SEA"

MONODONTIDAE THE ARCTIC WHALES

The monodonts take their family name from the singular tusk of the male narwhal ("mono" for "one" and "dont" for "tooth"). Both species in the family live exclusively in the Arctic, spending time in and around sea ice. The beluga is born dark, but as it ages its color turns to a distinctive pure white.

MAY WHISTLE TO EXPRESS JOY

BELUGA WHALE Delphinapterus leucas

ALSO CALLED "THE CANARY OF THE SEA" AFTER ITS BIRDLIKE SONGS

LARGE GROUPS OF BELUGAS ARE KNOWN TO CONGREGATE IN SHALLOW WATERS AND "EXFOLIATE" BY BEACH-RUBBING— LIKE A GIANT BELUGA SPA!

ON RARE OCCASIONS, MALE NARWHALS CAN GROW TWO TUSKS

It's hard to deny the mystique of the narwhal. The males of this species have a long, twisting tusk protruding from the upper lip, giving them a bizarre, whimsical appearance. It looks like some kind of horn or antler, but it is, in fact, made of dentin, the same material as teeth. Females rarely have a tusk. Very rarely, males can have a double tusk. Medieval Europeans believed the narwhal tusks they received from traders were unicorn horns. The people of the Arctic, though, knew the actual origins of the tusks, and have a long tradition of hunting the whales that continues to this day.

NARWHAL *Monodon monoceros*

UNUSUAL CONVEX TRAILING EDGES ON FLUKES

NICKNAMED "THE UNICORN OF THE SEA"

PHOCOENIDAE THE PORPOISES

The porpoises are sometimes referred to as "mereswine," which means "sea-pigs"! That term could be considered derogatory, but is actually quite fitting—pigs, like cetaceans, are among the smartest of all mammals. Porpoises are often shy, and typically live in smaller family groups than dolphins. All species in the porpoise family are "spade-toothed," with flattened, spade-shaped teeth (whereas dolphins tend to have pointy, conical teeth).

UNIQUE ANGULAR DORSAL FIN IS LOW ON THE BACK

BURMEISTER'S PORPOISE

Phocoena spinipinnis

SMALLEST—AND RAREST—SPECIES OF CETACEAN

VAQUITA

Phocoena sinus

DALL'S PORPOISE

NICKNAMED THE "PANDA OF THE SEA" AFTER ITS UNIQUE EYE MARKINGS

Phocoenoides dalli

THE FASTEST OF THE PORPOISES— AND FASTER THAN ANY OTHER SMALL CETACEAN!

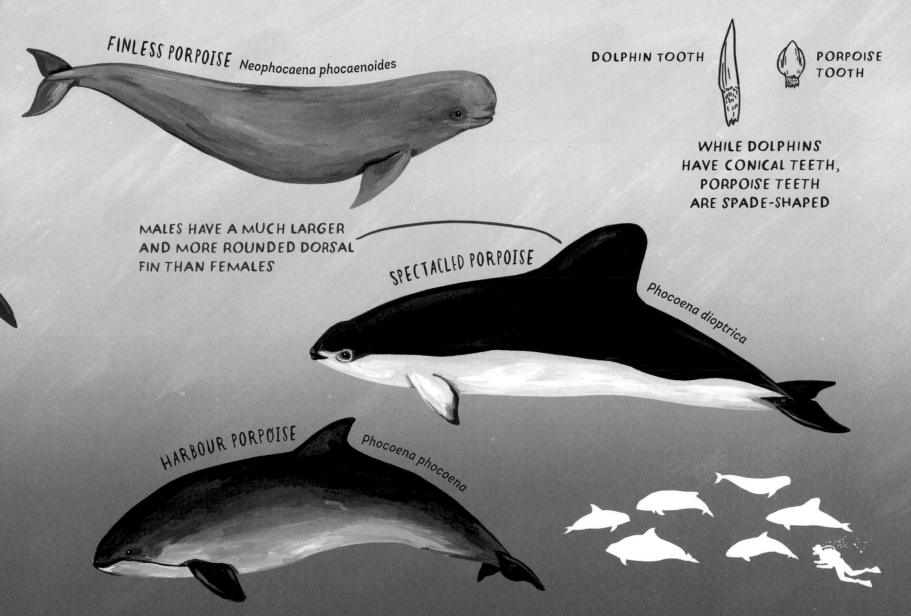

FINLESS PORPOISE *Neophocaena phocaenoides*

DOLPHIN TOOTH

PORPOISE TOOTH

WHILE DOLPHINS HAVE CONICAL TEETH, PORPOISE TEETH ARE SPADE-SHAPED

MALES HAVE A MUCH LARGER AND MORE ROUNDED DORSAL FIN THAN FEMALES

SPECTACLED PORPOISE *Phocoena dioptrica*

HARBOUR PORPOISE *Phocoena phocoena*

HUGE, SQUARED-OFF HEAD

SPERM WHALE *Physeter macrocephalus*

THE SPERM WHALE HAS THE LARGEST BRAIN OF ANY ANIMAL ON THE PLANET

IT IS THE LARGEST TOOTHED PREDATOR IN THE WORLD, AND AMONG THE LARGEST EVER TO EXIST

PHYSETERIDAE THE SPERM WHALE

The sperm whale gets its name from its spermaceti organ, which is unique to this species and its smaller counterparts, the pygmy and dwarf sperm whales. The organ contains waxy fluids that may be used to moderate acoustic signals and control buoyancy. Individuals of this species are very social, and female sperm whales are known to babysit infant whales whose mothers deep-dive for prey. The sperm whale, also known as the cachalot, is the largest of the odontocetes and was a major target of modern industrial whaling; it is still listed as a species vulnerable to endangerment.

64

KOGIIDAE THE LESSER SPERM WHALES

Like the true sperm whale, these species have a single blowhole angled off the left side of the head. Though the two smaller species are not as closely related to the true sperm whale as once believed, all three share a preference for the deep waters of the open ocean and have a blunted head shape, underslung lower jaw, and paddle-like flippers. The pygmy and dwarf sperm whales are sometimes mistaken for sharks, due to their size and shape and their "false gills" (white markings behind their eyes that look like the gills of a fish). These two species also have a unique, odd defense mechanism: expelling an inky, brownish liquid from their intestinal tract to cloud the water and confuse their predators.

WRINKLY SKIN

DWARF SPERM WHALE *Kogia sima*

FALSE GILL MARKING

POINTED DORSAL FIN, UNLIKE THE ROUNDED HUMP OF THE TRUE SPERM WHALE

PYGMY SPERM WHALE *Kogia breviceps*

PLATANISTIDAE THE SOUTH ASIAN RIVER DOLPHIN

This species comprises two main subspecies: the Ganges river dolphins and the Indus river dolphins, each found in and around the river systems of their respective names. Both subspecies have extremely small eyes that lack crystalline lenses, rendering them close to blind. They navigate not by sight but by echolocation. There is a great deal of color variation in these dolphins, which can be dark brown, light blue, or many shades of gray.

NICKNAMED "THE BLIND DOLPHIN"

SOUTH ASIAN RIVER DOLPHIN

Platanista gangetica

TEETH SHOW EVEN WHEN THE MOUTH IS CLOSED

PONTOPORIIDAE THE LA PLATA DOLPHIN

While other river dolphins live solely in freshwater environments, the La Plata dolphin is sometimes found in the sea along the eastern coast of South America. This dolphin, which also goes by the name Franciscana, has a similar body shape to other river dolphins and is shy and rarely seen in the wild.

LA PLATA DOLPHIN

Pontoporia blainvillei

LONGEST BEAK-TO-BODY RATIO OF ANY DOLPHIN—NEARLY 15 PERCENT OF ITS TOTAL BODY LENGTH

BOTO *Inia geoffrensis*

UPTURNED MOUTH CORNERS GIVE THE BOTO A SMILING APPEARANCE

INIIDAE THE SOUTH AMERICAN RIVER DOLPHIN

The South American river dolphin is well known for the adult male that can be colored a bright vivid pink. It's the largest of all the river dolphins, and local folklore tells of the dolphin leaving the river and shape-shifting into a man.

ZIPHIDAE THE BEAKED WHALES

The beaked whales are currently the most enigmatic of all cetaceans. Until recently, many species have been known only by their washed-up bodies and skeletons. Because cetacean bodies often change color soon after death, the true pigmentation of many species is unknown. Some beaked whales still have yet to be seen alive or at sea, and new species may be discovered or named in the future—here, twenty-two beaked whales are illustrated. Varying specimens have beaks of different sizes, and many have unusually configured teeth. They use echolocation to find prey, and a special adaptation allows them to hunt with suction, sucking squid and other organisms into their stomachs without biting them.

THE **LARGEST** OF THE BEAKED WHALES

BAIRD'S BEAKED WHALE *Berardius bairdii*

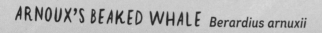

ARNOUX'S BEAKED WHALE *Berardius arnuxii*

BEAKED WHALES HAVE
"FLIPPER POCKETS"—
DEPRESSIONS ON THE
BODY FOR FLIPPERS TO
TUCK INTO

ABLE TO DIVE
FOR UP TO AN
HOUR AT A TIME

ZIPHIDAE THE BEAKED WHALES
CONTINUED

FOUND IN DEEP
INDO-PACIFIC WATERS

LONGMAN'S BEAKED WHALE *Indopacetus pacificus*

ALSO KNOWN AS
THE TROPICAL
BOTTLENOSE WHALE

UNIQUELY
BULBOUS
FOREHEAD

NORTHERN BOTTLENOSE WHALE *Hyperoodon ampullatus*

SOUTHERN BOTTLENOSE WHALE *Hyperoodon planifrons*

RANGES IN COLOR
FROM BLUEISH GRAY TO
BROWN OR YELLOW

KNOWN FOR REFUSING TO LEAVE ITS
WOUNDED COMPANIONS, WHICH
MADE IT VULNERABLE TO WHALERS

71

ZIPHIDAE THE BEAKED WHALES
CONTINUED

ANDREWS' BEAKED WHALE   *Mesoplodon bowdoini*

STRANGE,
STRONGLY
ARCHED
JAWLINE

HAS NEVER
BEEN SEEN IN
THE WILD

BLAINVILLE'S BEAKED WHALE   *Mesoplodon densirostris*

LARGE TEETH CAN
BECOME ENCRUSTED
WITH BARNACLES,
GIVING THEM THE
APPEARANCE OF
GIANT POM-POMS

SOWERBY'S BEAKED WHALE *Mesoplodon bidens*

MALES HAVE PROTRUDING TEETH AND ARE OFTEN SIGNIFICANTLY SCARRED, SEEMINGLY FROM COMPETITION WITH OTHER MALES

HUBBS' BEAKED WHALE *Mesoplodon carlhubbsi*

ZIPHIDAE THE BEAKED WHALES
CONTINUED

ALSO KNOWN AS THE
SKEW-BEAKED WHALE

HECTOR'S BEAKED WHALE *Mesoplodon hectori*

GRAY'S BEAKED WHALE *Mesoplodon grayi*

RECOGNIZABLE BY ITS
LONG WHITE BEAK, WHICH
IT RAISES ABOVE THE
SURFACE AFTER A DIVE

NAMED FOR THE ADULT MALES' TEETH, WHICH RESEMBLE THE LEAVES OF A GINKGO TREE

GINKGO-TOOTHED BEAKED WHALE *Mesoplodon ginkgodens*

AS IN MANY BEAKED WHALE SPECIES, THERE IS NO NOTCH IN ITS FLUKES

GERVAIS' BEAKED WHALE *Mesoplodon europaeus*

ZIPHIDAE THE BEAKED WHALES
CONTINUED

STRAP-TOOTHED WHALE

Mesoplodon layardii

PROTRUDING TEETH CAN
GROW TO BE 1 FOOT (30 CM)
LONG AND PREVENT MALES
FROM OPENING THEIR
MOUTHS FULLY

IT HAS NEVER BEEN
POSITIVELY IDENTIFIED
AT SEA AND IS KNOWN
ONLY FROM STRANDED
INDIVIDUALS

SIMILAR IN APPEARANCE TO
HECTOR'S BEAKED WHALE

PERRIN'S BEAKED WHALE *Mesoplodon perrini*

PYGMY BEAKED WHALE *Mesoplodon peruvianus*

THE SMALLEST OF THE BEAKED WHALES

TRUE'S BEAKED WHALE *Mesoplodon mirus*

THE MALES' TEETH ARE LOCATED RIGHT ON THE TIP OF THE JAW

77

ZIPHIDAE THE BEAKED WHALES
CONTINUED

Mesoplodon stejnegeri

STEJNEGER'S
BEAKED WHALE

ALSO CALLED THE
SABER-TOOTHED
BEAKED WHALE

DERANIYAGALA'S BEAKED WHALE Mesoplodon hotaula

THIS SPECIES HAS NEVER
BEEN OBSERVED AT SEA

SPADE-TOOTHED WHALE Mesoplodon traversii

SHEPHERD'S BEAKED WHALE *Tasmacetus shepherdi*

CUVIER'S BEAKED WHALE *Ziphius cavirostris*

ALSO CALLED THE
GOOSE-BEAKED WHALE

CAPABLE OF THE
DEEPEST DIVES OF
ANY CETACEAN

CHAPTER THREE
FOOD

As warm-blooded mammals—and oversized ones at that—cetaceans have high energy demands. Since they're constantly in motion, they burn a lot of calories just existing in their day-to-day lives and need lots of food.

Cetaceans are carnivores: animals that eat mostly other animals. Toothed whales hunt prey such as fish, squid, and crustaceans. Some species hunt larger prey—**killer whales**, for example, may eat sea mammals such as seals and sea lions, or even other cetaceans (hence their name). Baleen whales use their baleen to "filter feed," which allows them to feed on near-microscopic prey. The enormous **blue whale** feeds mostly on minuscule crustaceans called krill, and other baleen whales, which are generally very large, also feed on very small prey.

FILTER FEEDERS

All baleen whales use some method of filter feeding, but their specific styles of hunting vary from species to species.

MEALTIME FOR A GRAY WHALE

Gray whales are bottom-feeders, eating mainly small benthic creatures. They turn onto their right sides and swim along seabeds, sucking up sediment and filtering out prey as they go, often leaving a trail of mud in their wakes. Their baleen is usually worn down on the right side from rubbing against the sea floor, and the right sides of their faces can be scarred.

BENTHIC
Life forms at the bottom of a body of water.

SOME GRAY WHALES ARE "LEFT HANDED" AND FEED ON THEIR LEFT SIDE RATHER THAN THEIR RIGHT

MEALTIME FOR A RIGHT WHALE

Right and bowhead whales are "skim-feeders." They open their mouths and swim forward through the water, using their massive baleen to capture zooplankton in their mouth where they can easily swallow it. A bowhead whale's head can be a third of the size of its body length, and its baleen plates are the longest of any mysticete's. Their large, strong tail helps propel them forward against the drag caused by their massive head and baleen.

FILTER FEEDING

The whale shark, basking shark, and megamouth shark are all filter feeders, like baleen whales, straining their tiny prey out of the seawater rather than aggressively hunting larger prey. This, combined with their large size and their seemingly serene demeanors, makes them whale-like indeed. (And compared with some of the predatory toothed whales, they're downright tranquil!)

ZOOPLANKTON

A range of tiny organisms that drift through ocean water.

ONE DIVE CAN YIELD
HALF A MILLION
CALORIES-WORTH
OF FOOD

A BLUE WHALE CAN
EAT 40 MILLION KRILL
IN A SINGLE DAY

MEALTIME FOR A RORQUAL

The rorquals—the whales of the family *Balaenopteridae*—have
pleated grooves that enable their throats to expand massively.
This allows the rorqual to take in a tremendous mouthful of
water—sometimes matching the volume of the whale itself—
before filtering it back out, trapping the tiny prey in its mouth.

KRILL

THE HUGE CLOUDS OF KRILL THAT WHALES FEED ON ARE MADE UP OF TINY INDIVIDUAL CRUSTACEANS

OPPORTUNIST BIRDS

It's common to see gatherings of seabirds on the surface of a whale feeding session (or, if they're diving birds, below the surface), since birds and whales often eat a lot of the same foods. As a result, the presence of some types of birds can signify to whale watchers that a certain whale species is in the area. Phalaropes, for example, eat lots of krill—so a group of these birds on the surface of the ocean may signal that a blue whale is also in the area, pursuing the same food source.

HUMPBACKS BLOW DENSE WALLS OF
BUBBLES TO CONFUSE THEIR PREY

BUBBLE NETTING

Humpback whales are famous for a hunting technique called "bubble netting," in which they work together to herd shoals of fish into a tight "bait ball" for easier hunting. It's a brilliant strategy, because small fish often swarm tightly together as a defense mechanism against predators—the whales use this to their advantage.

A bubble netting starts when an individual whale begins to blow a wall of bubbles around a shoal of prey fish such as herring or capelin. While the bubbles confuse and startle the fish, the other whales make sounds that may help each other coordinate movements and further dissuade the prey fish from escaping the bubble net. Once the fish are sufficiently gathered together near the surface at the water, the whales lunge upward in a coordinated feeding formation.

HOW TOOTHED WHALES EAT

Odontocetes do not chew; instead, they swallow their prey whole. They're called "toothed whales," but not all odontocete species have many teeth. Narwhals, for example, have only one tooth—their tusk—and since only males typically have tusks, most female narwhals are functionally toothless. Similarly, among many of the beaked whale species, only males have teeth, which are likely used for fighting among males rather than for eating.

BEAKED WHALES HAVE THROAT GROOVES THAT ENABLE THEM TO DROP THE PRESSURE IN THEIR MOUTH AND CREATE A TEMPORARY VACUUM, ALLOWING THEM TO SUCK IN PREY

OTHER ECHOLOCATORS

Bats, the flying mammal relatives of cetaceans, and swiftlets, a type of bird, also use echolocation to find their prey, though their techniques evolved separately.

ECHOLOCATION

Toothed whales have a special adaptation that distinguishes them from their mysticete cousins: the ability to echolocate.

A specialized organ called the "melon" in the front of a toothed whale's head can project sound outward into the surrounding environment. When the echoes bounce back to the cetacean's head, it can interpret the information and use it to navigate and to hunt prey.

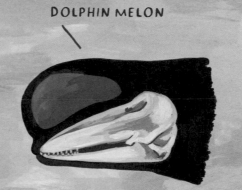

DOLPHIN MELON

BELUGA MELON

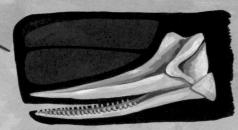

SPERMACETI ORGAN

SPERMACETI

The sperm whale's head houses its namesake—the large structure known as the spermaceti organ, which aids in sound production for echolocation. Spermaceti oil harvested from dead whales was a sought-after product during the height of commercial whaling.

HERDING

Odontocetes have a number of methods for herding fish into ideal formations for hunting. The common dolphin can herd sardines by creating walls of bubbles to concentrate them, in a manner similar to the bubble netting used by humpback whales. Other techniques include using tail-slapping to startle and herd fish, chasing shoals of fish and cornering them against natural barriers such as beaches, and even tracking fish using echolocation.

MUD RINGS

Off the coast of Florida, bottlenose dolphins use their tails to stir up silt from the sea floor, creating a "mud ring" around a school of fish. This disorients and traps the fish, who attempt to jump up and out of the mud ring, where the dolphins catch and eat them.

SPECIALIZED HUNTING

A diverse array of hunting methods have been observed among the toothed whales.

Among orcas, different pods eat different prey; some pods hunt fish exclusively, while others will also hunt smaller sea mammals such as dolphins or seals. Different pods also have different hunting strategies, passed down from generation to generation within the pod but not taught to outsiders. Orca pods are so highly individualized that they often even "speak" completely unique languages, with calls, clicks, and whistles distinct to each pod. They also earn the nickname "killer whale" with their somewhat heartless-seeming approach to hunting other animals: they'll often eat only a nutrient-rich internal organ or two and let the rest of the carcass drift away. Orcas challenge the notion of whales as peaceful, gentle giants.

Narwhals hunt by sucking prey into their mouths; but recently, male narwhals were filmed using their tusks to whack prey, stunning it and making it easier to catch. (Previously, the use of the narwhal tusk was a mystery, but now we have evidence of at least one of its uses.)

HUMAN-CETACEAN COOPERATION

Cetaceans are incredibly intelligent creatures, capable of discovering new techniques and strategies for survival in the wild and passing that knowledge down to their offspring. Learned knowledge is passed through generations of cetaceans in a way that's not unlike how we humans share our knowledge with one another. In addition, cetaceans and humans actually can cooperate in some really interesting ways. In Laguna, Brazil, murky water can make it difficult for net fisherman to know where and when to cast for mullet. A nearby population of bottlenose dolphins—who can locate fish via echolocation, despite the murky water—has learned to herd the mullet toward the shore, diving to communicate to humans on the beach that their catch is now within reach. The dolphins might be taking advantage of the added chaos of the cast nets, which could make the mullet a quicker catch.

SPONGING

Some dolphins use sea sponges as a kind of cap on their beaks as they rummage the sea floor to find food, possibly to protect their beaks from getting scratched up by rocks and other sea debris. It's believed that this sponging technique is handed down through dolphin generations by mothers, who teach it to their calves; its origins may stretch back hundreds of years.

SPERM WHALES CAN DIVE
HUNDREDS OF METERS
BELOW THE SURFACE

SPERM WHALES VERSUS GIANT SQUIDS

For hundreds of years, sailors circulated rumors of an enormous tentacled sea monster living in the ocean depths. For a long time, there was no official record of such an animal, and its status remained mostly legendary until widespread sperm whale hunting revealed hard evidence of this underwater colossus. Captured sperm whales often bore large circular scars, and in their stomachs, human hunters found mysterious beaks—like the hooked beak of a bird, but far larger. These turned out to be the beaks of the giant squid—discovered courtesy of its only predator.

The battle between the huge squid and the even bigger whale is one of the most colossal confrontations between two animals on Earth.

Eventually, the bodies of some giant squid were discovered, confirming the species' existence. And the species has now even been observed alive. But to this day, the epic battle between a sperm whale and a giant squid has never been caught on camera.

SAILORS CALLED THE MYSTERIOUS CREATURE THE "KRAKEN"

WHALE FALLS

Occasionally, when a whale dies, its carcass falls to the ocean floor and becomes the basis of a unique and complex kind of ecosystem called a "whale fall." Whale falls are so deep and remote that humans didn't discover and study them until the 1980s, and it was a stunning scientific discovery indeed. It turned out that whale falls were home to species that had never been seen before, and that the way the whales are digested by different animals and bacteria creates a specialized ecosystem for the deepest of sea creatures that can last for decades.

POLYCHAETE WORMS

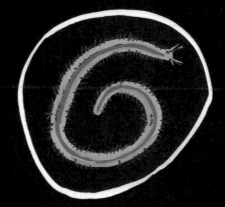

GIANT ISOPODS

HAGFISH

SEA CUCUMBERS

SQUAT LOBSTERS

OTHER CREATURES SEEN
AT A WHALE FALL:

- SLEEPER SHARKS

- PRAWNS

- SHRIMP

- CRABS

LITTLE COLONIES

Whales are often hosts to entire populations of parasitic (or quasi-parasitic) guests.

REMORAS

Smaller dolphins and porpoises are targets of parasites too. Remoras, a kind of ray-finned fish, have sucker-like heads that allow them to attach to cetacean hosts of all sizes, which offers the remoras protection as they swim. They clamp on to many large animals, including sharks, sea turtles, and dolphins. Because they don't harm their hosts, remoras are not technically parasites—but they may annoy dolphins, which may try to scrape remoras off their skin.

WHALE LICE

Whale lice are small crustaceans that live on the bodies of cetaceans of all shapes and sizes. They infest folds in the skin, wounds, and any other openings where they can catch hold. Perhaps the most notable are the whale lice that congregate on right whale callosities. The callosities' white color actually comes from the whale lice and other parasites that live on them.

WHALE LICE CAN BE UP TO 1 INCH (2.5 CM) LONG

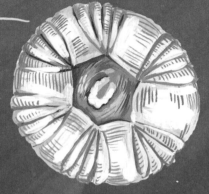

BARNACLES LIKE THIS ONE GROW DIRECTLY ON THE BODIES OF WHALES

BARNACLES

A barnacle is a type of crustacean that attaches itself permanently to a solid surface. Whale barnacles live on the skin of cetaceans. Some are considered parasites; others are considered benign. They are a common home for whale lice.

CHAPTER FOUR

HABITATS

More than 70 percent of the earth's surface is covered in oceans, and cetaceans can be found in all of them, from warm tropical waters and muddy freshwater rivers to icy polar seas.

The variety of habitats occupied by cetaceans is a testament to the spectacular diversity among these species. From deep divers whose anatomy has adapted to withstand extreme underwater pressure to river dwellers who have evolved extra-flexible bodies for navigating twists and turns, each cetacean species is adapted to its surroundings in a unique way. This highlights how important those precious habitats are to cetaceans' survival. In this chapter are just a few of the many environments occupied by cetaceans.

MIGRATION

Whales are known for extremely long seasonal migrations, with some species traveling farther than any other non-human mammals to take advantage of varied habitats that serve varied needs. For example, many humpback whales breed and give birth to their young in the warm waters of the tropics—an ideal environment for new babies—but migrate seasonally to cold polar seas, where food is more plentiful.

TUSKING

Male narwhals are sometimes observed crossing tusks with each other above the surface of the water; this behavior is known as "tusking." In the past, this was believed to be an aggressive behavior, but researchers now believe it is a friendly social behavior.

ARCTIC

The Arctic—the northernmost polar region of the globe—is home to many cetacean species for part of the year, while others live there year-round. Although belugas migrate hundreds of miles between their summer and winter habitats, they don't ever leave the cold waters of the north. Bowhead whales spend their lives in the Arctic; their extremely thick blubber—up to 1.5 feet (0.45 meters) thick—helps them withstand colder temperatures. Narwhals also live exclusively in the Arctic, using cracks in sea ice as passageways for their migrations across Arctic waters.

Cetaceans such as the killer whale, Arnoux's beaked whale, humpback whale, and Antarctic minke whale spend time in Antarctic waters at the globe's South Pole.

OCEAN ACIDIFICATION

Even though they occupy only less than 1 percent of the
sea floor, coral reefs house 25 percent of all ocean life. A
major threat currently facing coral reefs is the acidification
of ocean water. As the level of carbon dioxide (CO_2) in the
earth's atmosphere rises, the oceans also absorb more of
that CO_2, lowering the pH of seawater and making it more
acidic. These conditions make shell growth difficult for many
organisms, including coral, and we are in danger of losing
many important reefs altogether.

CORAL REEFS

Found in warmer tropical waters, coral reefs are an important habitat for many living creatures. Some people call coral reefs the "rainforests of the sea" because of the incredible diversity of life they support. When picturing coral reefs, we often think of smaller life forms like coral, anemones, and brightly colored fish, but coral reefs are also important feeding grounds for cetaceans. Many cetacean species visit the Great Barrier Reef of Australia, the largest coral reef system in the world. Humpback whales use the Great Barrier Reef as a breeding ground. Dolphins are often found in and around coral reef systems, too.

COASTLINES

Porpoises tend to favor the shallow waters along coastlines, which has made them more vulnerable to coastal fishing, pollution, toxic runoff, and boats. Bottlenose dolphins are a charismatic species that often lives near shorelines. Coastal lagoons are also popular territory for cetaceans, and the San Ignacio Lagoon in Mexico has long been a winter sanctuary for mother gray whales and their calves. In the past, whalers exploited the high density of gray whales that needed the lagoon, but these days, thanks to conservation work, it is better known for whale watching and ecotourism.

THE DEEPEST-DIVING WHALES TRAVEL TO SUCH EXTREME DEPTHS
THAT THEIR LUNGS ARE ACTUALLY ADAPTED TO COLLAPSE
UNDER THE HIGH PRESSURE

OPEN OCEAN

Cetaceans that spend most of their time in the open ocean are often mysterious because their large habitats and wide distribution make them difficult to observe. The beaked whales of the family *Ziphidae*, for example, almost always live in the deep ocean. In fact, they're some of the deepest divers in the animal kingdom. Their remote nature has historically protected them from humans and they have almost never been targeted by whalers, since they spend most of their time in remote reaches of the sea. Changing oceans as a result of climate change, however, will impact their lives.

THE TUCUXI, THE SOUTH ASIAN RIVER DOLPHIN, AND THE FRANCISCANA LIVE IN RIVERS AND COASTAL ESTUARIES

RIVERS

Though most cetaceans are truly marine, spending most or all of their lives in seawater, a few dolphin species live in fresh water. The boto, also known as the pink dolphin, lives in the Amazon and is specially adapted for the tighter movement constrictions of life in the river. Their cervical vertebrae are not fused, meaning the boto can bend its neck and turn its head from side to side. This, along with the botos' well-adapted flippers, makes their bodies much more nimble than their ocean-dwelling counterparts, so they can more easily navigate the shallows of their habitat.

VARVARA THE GRAY WHALE
WAS TRACKED MAKING
AN ASTONISHINGLY LONG
MIGRATION

MIGRATION

While some cetacean species keep within a small geographic range no matter the time of year, other species are known for some of the longest migrations in the animal kingdom. Scientists are able to study the migrations of various species using tracking devices fitted to individual animals, as well as analysis of photographic records of individual whales. Many humpback whales migrate seasonally from high latitudes, where they feed, to lower latitudes, where they give birth, and for a long time, they were thought to have the longest migration of any mammal. But in 2011, the humpback's record was smashed when a gray whale named Varvara who had been fitted with a satellite tag was tracked completing a round trip migration totaling 14,000 miles (22,530 km).

CHAPTER FIVE

FAMILY, LIFE, AND SOCIETY

Cetaceans have unique and complex social lives, much of which we are only just beginning to understand. But in recent years, we've become more acquainted with the inner workings of cetacean society, which can be sophisticated, complex, and fascinating.

As humans, we tend to believe that we exist apart and above other life on Earth—that we, as critical thinking, free willed, emotional beings, are smarter than any other species. This belief in our own superiority is called **human exceptionalism**. In some ways it makes sense. After all, we're the species whose population has exploded into every corner of the earth, and whose actions have incredibly wide consequences for all the planet's lifeforms. However, when we look closely at the lives of cetaceans, human exceptionalism starts to feel a little ridiculous.

Familiar pieces of our lives are mirrored in those of our cetacean cousins—when they babysit their sisters' young calves; when they perform seemingly altruistic acts, even on behalf of other species; when they invent new hunting methods and pass them down through the generations. The more we learn about cetaceans, the more our differences seem to dissolve away, along with the notion that we alone are the only intelligent, compassionate beings on the planet.

MATING

Cetaceans usually mate with many partners throughout a breeding season. Some species are competitive. Humpback courtship often involves a lengthy period of fierce competition during which several males pursue a female through the water, lunging at each other in the process, using their tails to bash neighboring males, and even drawing blood—sometimes the rivalry can become so intense that males are killed in their pursuit. Other species are more cooperative; some male dolphins form friendship bonds with other males and act as a lookout for each other during mating, helping to ensure that no rival males can mate with a given female.

SEXUAL DIMORPHISM

"Sexual dimorphism" is the term used when males and females of a species look different from one another. This is often due to sexual selection—the species' preference for certain characteristics in their mates. Many cetaceans exhibit sexual dimorphism. In sperm whales, for example, males can be three times the weight of females, and have much more pronounced square-shaped heads. Many beaked whale species have different jawlines and dental configurations between the sexes, and male narwhals have a tusk, while females do not.

FEMALE NARWHALS ARE USUALLY TUSKLESS,
UNLIKE MALE NARWHALS

MALE KILLER WHALES
HAVE LARGER AND TALLER
DORSAL FINS

MALE HUBBS'
BEAKED WHALES
HAVE WHITE "HAT"
COLORATION ON
THEIR HEADS, AS WELL AS
LARGE VISIBLE TEETH

117

GESTATION, BIRTH, AND INFANCY

Like other highly developed species, cetaceans spend a long time in the womb. This period is known as "gestation," and while its length varies among cetacean species, it is always long relative to that of most other mammals.

Cetaceans almost always give birth to one baby at a time. While land mammals typically deliver their young head first, cetaceans deliver tail first, to provide the newborn with the ideal positioning to take its first breath. Depending on the species, the mother or another adult may help the baby to the surface to breathe.

All cetaceans nurse their young, and their milk is much richer and fattier than the milk of land mammals. Its thicker texture helps calves drink more without losing milk to the surrounding water.

CETACEAN CALVES CAN RIDE IN THEIR MOTHER'S SLIPSTREAM, WHICH ALLOWS THEM TO REST AND EVEN SUCKLE WHILE TRAVELING

NARWHAL CALVES ARE BORN
WITHOUT TUSKS AND HAVE A
GRAYER, MORE EVEN SKIN TONE
THAN THEIR PARENTS

YOUTH AND BEYOND

The juvenile period for cetacean calves lasts as long as their mother is nursing them, which is generally about a year, though some species go longer. Sperm whales, for example, may nurse their young for many years. When the young are no longer dependent on their mothers for nutrition and can feed independently, they are considered adolescents. Some species leave their birth families and go off on their own, but some stay with their pods the rest of their lives. A cetacean is categorized as a true adult when it reaches sexual maturity, the age at which its body is capable of creating new life. This age also varies widely across cetacean species.

When sperm whale calves are very young, they aren't capable of deep diving. But their mothers need food to sustain themselves and the milk they are producing for their offspring. Sperm whale mothers are known to leave their babies at the surface of the water, often with "babysitter" whales—related female adult sperm whales—who watch and guard the calves while the mothers carry out necessary deep dives for food.

SONG AND SOUNDS

While echolocation is critical to many cetaceans' hunting strategies, many also use sound to communicate with each other. Some make noises while deep underwater, while others vocalize above the surface. Sound can enhance cetacean mating behaviors, help mothers and calves stay together, and even help groups of individuals coordinate their movements.

HUMPBACK SONGS

The songs sung by male humpbacks are the longest and most complex of any nonhuman songs in the animal kingdom. They are composed of sequences of five to fifteen phrases that are repeated for up to several hours. Humpbacks can use their song to communicate with whales miles away, and different populations of humpbacks have different singing styles. We're still not totally sure why humpbacks sing, though it's suspected that their songs have some role in courtship and mating.

BELUGA VOCALIZATIONS

One of the most vocal cetacean species, belugas communicate with a large repertoire of squeaks, whistles, chirps, groans, clicks, and other sounds. Beluga calves learn to communicate with sound from a young age and use unique sounds to communicate with their mothers.

THE BOOMING BLUE WHALE

Blue whales make sounds with the highest decibel levels of any animal, which can be heard up to 500 miles (805 km) away.

INTERSPECIES INTERACTION

Cetaceans are often observed interacting with each other across species—dolphins of different species may travel together for a while, for instance. Humpback whales in particular, though, are known for their friendly behavior toward non-humpbacks. They have been seen playing with bottlenose dolphins, lifting them out of the water with their heads, and protecting gray whale calves from being attacked by killer whales. And it's not just other cetaceans they interact with—humpbacks are also known to help seals, sea lions, and even large fish. One humpback whale was seen rolling onto its back to carry a Weddell seal to safety after it had been knocked off its ice floe by killer whales.

FUN AND PLAY

Cetaceans aren't just some of the most intelligent animals on the planet; they're also very playful. Dolphins are particularly famous for their recreational activities; they often play with "toys" in the form of bits of seaweed or other debris, picking them up with their mouths and tossing them around. They also produce "bubble rings," donut-shaped rings of air that they play with. Dolphins are one of the few examples in nature of animals intentionally becoming intoxicated: they have been observed biting pufferfish—seemingly intentionally—causing the fish to release small amounts of a neurotoxin that puts the dolphins in a trance-like state.

SLEEP

Most mammals breathe automatically, without conscious effort. But because cetaceans live in the water, they must swim to the surface to breathe. This makes sleeping complicated, and different cetaceans handle sleep differently. Many species fall asleep with only one side of their brain at a time—while that half snoozes, the other half monitors their surroundings and reminds them to surface for a breath. Then they'll switch sides and let the active side sleep.

SPERM WHALES SLEEP VERY LITTLE, BUT THEY HAVE BEEN OBSERVED NAPPING IN THIS SURPRISING FORMATION—HANGING VERTICALLY TOGETHER SEVERAL METERS UNDER THE SURFACE

CHAPTER SIX

HUMANS

The relationship between cetaceans and humans is long and storied, evidenced by their presence in our folklore, scripture, traditions, and art. We've long been fascinated by our sea-dwelling mammalian cousins, which has had both positive and negative implications for cetaceans themselves. While first regarded as a sacred resource by some, whales unfortunately fell victim to exploitation by humans through the peak of the commercial whaling industry in the nineteenth century. Thankfully, a change occurred in the mid-twentieth century, and attitudes in many parts of the world began to shift toward treating cetaceans as precious and worth protecting, rather than abusing them. But even now, our relationship with our underwater kin continues to evolve, and our actions as humans, even when we might not realize it, have an impact on them. We're learning more every day about how special and important these beautiful creatures are, and what they need from us to survive, but our human systems can be slow to change. While we have a long way to go to protect cetaceans, we can make a difference by learning from our past as we look to our future.

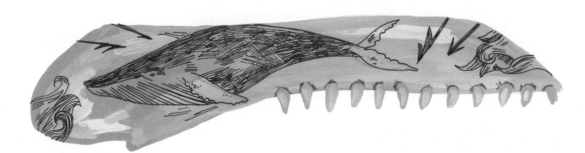

CETACEANS IN MYTHOLOGY

For much of human history, we could only guess as to the inner workings and lives of cetaceans. For many cultures throughout history, even catching a glimpse of a whale up close was a rare, brief, and special occurrence. Inevitably, whales became the source of myths and legends among humans who were aware of them, but didn't know exactly what they were or have a full understanding of them (they were often mistaken for sea monsters). For many other cultures, whales have always been a common sight and even a familiar food source. These seafaring and coastal cultures often have long, rich traditions of storytelling and myths deeply connected with the mammals of the sea.

CETACEAN CONSTELLATIONS

There are cetaceans in our maps of the cosmos in the form of Delphinus, whose name is Latin for dolphin, and Cetus, who is named for a mythical sea monster. Cetus also shares etymological roots with the term "cetacea" and is often depicted as a whale in modern star maps.

SEDNA

In Inuit mythology, the goddess Sedna's fingers were chopped off; they turned into walruses, seals, and whales. A possible dwarf planet at the outer reaches of our solar system is named after this figure.

NATSILANE

In Tlingit mythology, the hunter Natsilane carved a blackfish out of wood and threw it into the sea, where it became a living whale.

HROSSHVALUR

In Icelandic mythology, the Hrosshvalur was a fierce whale with a horse's head and a long red mane.

MAKARA

In Hindu mythology, the sea creature Makara is sometimes depicted as half land-animal, half dolphin.

PRINCESS BAIJI

In one Chinese legend, the baiji river dolphin (which is now considered functionally extinct) was a transformed princess who had been cast into the Yangtze river.

PAIKEA

In Maori legend, the ancestor Paikea rode to New Zealand on the back of a whale.

CETACEANS IN ART AND DESIGN

Whalers' carvings using the teeth, bones, and baleen of whales are known as scrimshaw. The craft was developed on board whaling vessels, where sailors needed a way to pass the time—they used the materials they had at hand as their canvasses. Baleen was used in the basket-making traditions of the Kinguktuk and others. Dried and then reconstituted in water, baleen could be cut into strips perfectly suited to precise basketry. Carved whale bone is a traditional example of a *taonga*, or a cultural treasure, in Maori society.

Environmentalist Roger Payne recorded and released an album in 1970 entitled *Songs of the Humpback Whale*, which is sometimes credited with helping to change public attitudes about the need for whale conservation. Whale song has even traveled into interstellar space. When the two Voyager missions launched from Earth to travel to the outer reaches of the solar system and beyond, they carried some unusual cargo: engraved golden phonograph records that contained sounds and song from all around the world. One of the tracks—nestled among the sounds of thunderstorms, human voices, and chirping crickets—is the song of a humpback whale.

The novel *Moby-Dick; or, The Whale* by Herman Melville is famous for its opening line, "Call me Ishmael." The American classic follows a whaler, Captain Ahab, on his quest for vengeance against a white sperm whale who had bitten off his leg. The white whale in the story was partly inspired by a real-life albino sperm whale that whalers called Mocha Dick, and Ahab's crusade inspired the colloquial use of the phrase "white whale" to describe something that someone seeks that is always just out of reach.

KILLER WHALE JAWBONE

WALRUS TUSKS WERE ANOTHER
POPULAR CANVAS FOR SCRIMSHAW

SPERM WHALE TEETH

133

AMBERGRIS

Ambergris is a smelly, waxy substance formed in the digestive system of some sperm whales. Its function is thought to be protective: when hard, sharp objects like squid beaks pass through the whale's intestines, they become covered in ambergris, which may stop them from damaging the whale's internal organs. The whales pass lumps of ambergris either through vomiting or in fecal matter, after which the waxy substance floats to the ocean surface.

Despite its slightly unpleasant origins and the fact that it has an unpleasant smell when fresh, it was once a very popular and expensive ingredient in high-end perfumes. Since it is naturally expelled, it can be harvested without harming a whale. But as a by-product of a protected species, it is illegal to sell and use in many areas of the world. Because of this and the advent of new synthetic perfume ingredients, ambergris has greatly fallen out of favor in perfume-making.

AMBERGRIS

SQUID BEAK

UNICORNS OF THE SEA

Europeans of the Middle Ages and early modern period believed unicorn horns had magical properties—including the ability to detect and neutralize poisons—so they were sought after by politicians and other notable figures. Queen Elizabeth I famously bought one of these "unicorn horns," which was actually a narwhal tusk. At the time, the general public and even naturalists in Western Europe believed that unicorns were real—if very rare—animals, and not creatures of myth. Explorers, traders, and people who lived near narwhals were familiar with this anatomical curiosity. But those who had never seen or even heard of narwhals, can be forgiven for thinking the tusks had mythical origins.

NEW OBSERVATIONS

The lives of many ocean creatures have long been mysterious to humans. The ocean is vast, and cetaceans have many behaviors that humans just can't observe that easily. But technology is always improving, with new tools and methods being invented. It's an exciting time in cetacean observation, because drones—small, remote-controlled aircraft—have become smaller and quieter, and the technology that allows them to take high-quality photos and videos has improved. Drones are able to track and visually capture cetaceans in their natural habitats with minimum disruption (planes and helicopters are too loud for this purpose). Often, a whale might not even know a drone is flying above it.

Exciting drone footage has captured a number of incredible cetacean behaviors in recent years, offering proof of some theories about cetacean life while also uncovering incredible, unexpected evidence of behaviors. In addition to being an excellent new source of data for scientists studying marine life, these videos help us land-bound humans get a glimpse into our cetacean cousins' lives and gain some appreciation for their majesty at sea.

DOLPHIN STAMPEDE

Dolphins often travel in pods of about a dozen individuals, but occasionally many pods come together to form a "superpod" that can be several miles across. Drones off the coast of California have observed such dolphin "stampedes"—thousands of common dolphins, porpoising together in the same direction.

NARWHAL TUSKS

The purpose of narwhal tusks has stumped humans for centuries. Only males have tusks, so they may serve some evolutionary purpose in competition for mates. But do the tusks have other uses? Recently, drones have captured footage of narwhals smacking fish with their tusks to stun them before sucking them into their mouths. So there may be a function of the narwhal tusk that we were previously unaware of: a hunting tool.

LUNGE FEEDING

Even though we've known about lunge feeding for years, we had yet to get a good overhead view of the behavior. Now, thanks to drones, we have detailed overhead footage of multiple rorqual species lunge feeding on clouds of krill and plankton.

WHALING

The practice of hunting whales is called whaling. Many indigenous societies have hunted whales for centuries, but it wasn't until the advent of modern commercial whaling that humans became such a massive threat to cetacean species around the globe.

Subsistence whaling is an important part of a number of indigenous societies worldwide. As a result, conservationists sometimes blame such cultures for contributing to the crisis of decline in cetacean species and populations. But even while native cultures have practiced whaling for far longer than the era of commercial whaling, the lasting harm of commercial whaling far eclipses any harm inflicted by native groups. The scale of commercial whaling was just too enormous, and was carried out without concern for the species it exploited. The International Whaling Commission (IWC) issued a "commercial whaling moratorium" that banned commercial hunting of all cetaceans. While most commercial whaling has ended, Norway, Iceland, and Japan still allow the practice, and many cetacean species are under threat of endangerment as a result.

The IWC regulates whaling by indigenous groups in some countries, and whaling in most countries is allowed only on a subsistence basis and only by native people.

FISH OR MAMMALS?

Today, we know that cetaceans are our mammal relatives. But not so long ago, there was still fierce debate over whether they were mammals or . . . fish. In 1818, a court trial over whale-oil regulation resulted in a ruling declaring that whales, who lived in the sea and didn't have feet, must be fish. Even though scientists had considered whales mammals for years, the general public hadn't fully grasped this fact. Perhaps our unwillingness to see ourselves in our cetacean relatives was tied to our greed in exploiting them as an economic resource.

BALEEN AND OIL

In the heyday of commercial whaling, people found countless commercial uses for baleen. At the time, baleen was referred to as "whalebone," and whalers found that baleen—a flexible and strong material in a preplastic world—could be a profitable commodity alongside whale meat and whale oil. As commercial whaling grew, commercial uses for baleen started to pop up, though indigenous groups had long used baleen in traditional applications such as basketry.

WHALE MEMORY

Some whales are old enough to have been alive during the peak of industrial whaling. Bowhead whales, for example, can live to be more than two hundred years old and may remember the not-so-distant days when human whalers hunted them.

STILL WHALING

An IWC subcommittee oversees aboriginal subsistence whaling, which sets different quotas for whales that can be taken by aboriginal groups.

FROM WHALING TO WHALE WATCHING

Today when boats seek out whales, it's more likely to be for the purpose of whale watching. If you have the chance to go on a whale-watching excursion, you might get to see one of nature's giants up close and personal. Make sure to do your research ahead of time and go with a credible institution—your captain should know how to keep a distance that's safe for both you and the cetaceans. It's also nice to have a naturalist on board who can help spot species and provide context and teaching along the way.

Because whales, dolphins, and porpoises live throughout the world's oceans, whale watching can take place almost anywhere near the sea. The best times to whale watch depend on where you are and the migration cycles for local whales. Some particularly exciting whale-watching sites include the Bay of Fundy in Canada, which attracts many species of whales, such as the endangered North Atlantic right whale; Baja California in Mexico, a destination of many gray whales' record-breaking migrations; and Disco Bay in Greenland, where the many species of local cetaceans include the elusive narwhal.

WHALE WATCHING GEAR

Here are a few items that are good to bring along when you go whale watching.

BINOCULARS
Even at close range, binoculars can help whale watchers see whales in better detail.

WATERPROOF BAG, WINDBREAKER/RAINCOAT
Boat trips can get wet. Wear waterproof clothing to protect yourself, and rubber-soled shoes to help you keep your footing on a slippery boat deck.

HAT AND SUNGLASSES
There's no shade on the open ocean, so consider taking added precautions to protect yourself from the sun. Polarized sunglasses can also help reduce glare in your vision, making it easier to spot whales.

THE MOST VULNERABLE SPECIES

VAQUITA

After the baiji was declared extinct, the vaquita took the title of the rarest cetacean. Now the little vaquita—which is also the smallest cetacean species—is on the brink of extinction, too. Their population had fallen to around a hundred individuals by 2014, and as of 2017, that number had fallen to an estimated thirty individuals— a critically low number. The main threat to the vaquita, which is endemic to the Gulf of California in Mexico, is illegal fishing. Gillnetting—a serious threat to the little porpoise—was permanently banned by the Mexican government in 2017, but with the vaquita's numbers already perilously low, some conservationists worry that it's too late to save the species.

THE BAIJI

Until recently, no cetacean had gone extinct in modern times. But in 2006, the baiji (also known as the Yangtze river dolphin and the only member of the family *Liptidae*) was declared functionally extinct after an expedition of scientists traveled for six weeks searching for individuals of the species and was unable to find any. None have been seen since, and it is very likely that the baiji is gone forever. The primary cause of its demise was the industrialization of the world around it. As boats crowded the Yangtze river, the water was polluted—not just by chemicals, but by sound as well. For an animal that relied on sound to navigate and communicate with fellow baijis, this proved a fatal blow to their survival.

OTHER VULNERABLE SPECIES

Besides the critically endangered vaquita, the International Union for the Conservation of Nature (IUCN) lists seven cetacean species as endangered: the blue whale, the fin whale, the sei whale, the North Atlantic and North Pacific right whales, the South Asian river dolphin, and the Hector's dolphin. Several other species are categorized as "near-threatened," which means they are at risk of becoming endangered in the near future.

Though whaling has been outlawed in most countries, there are a number of other serious threats to cetaceans throughout the world. Fishing practices are one, as cetaceans are mistakenly caught in nets as by-catch. Collisions with large shipping vessels can kill cetaceans, and oil spills and other toxic contamination threaten the health of all marine animals. Overuse and overpopulation by humans in areas that cetaceans frequent can also threaten their health and safety; human activity continues to encroach on marine areas that whales, dolphins, and porpoises need in order to maintain healthy populations. Even sound pollution from ships is a threat to species that rely on echolocation.

Possibly most threatening and hardest to predict is human-caused climate change. Warming sea temperatures mean changing ecology in the polar habitats of many cetaceans. Ocean acidification as a result of increased CO_2 levels in the atmosphere is causing habitat degradation and has the potential to deplete food sources for many cetaceans.

CETACEANS IN CAPTIVITY

Most cetacean species cannot be successfully kept in captivity. They need the freedom to swim long distances and make deep dives, though some species—including orcas, belugas, and bottlenose dolphins—are currently kept in zoos, aquariums, theme parks, and rescue facilities around the world. Public attitudes toward keeping cetaceans in captivity—especially for the purpose of human entertainment—are changing. Aside from the few individuals that are injured or sick and can survive only in rescue facilities, cetaceans are increasingly thought of as wise, complex creatures that not only need, but also wholly deserve, wild space to live and thrive.

CONCLUSION

The last few hundred years have seen the devastation of many cetacean species, but the story doesn't end here. As we've become more aware of the challenges faced by cetaceans at the hands of humans, we've also learned more about how we can help them. Many species that were pushed to the brink of extinction during the commercial whaling boom have seen incredible recovery in their population numbers due to conservation efforts like whaling restrictions and habitat protection. Even as we are capable of destroying species forever, we can also protect them. The world is home to all of us, and if we can learn to remember our underwater cousins and honor their lives in the oceans and waterways of the world, we can allow cetaceans to live and thrive for generations to come.

HOW TO HELP WHALES

I hope that this book has inspired appreciation for our majestic cetacean relatives. If you want to make a difference for these species, here are a few things you can do.

STAY UPDATED

Follow Instagram accounts like those of wildlife photographers @paulnicklen and @joelsartore to stay updated on conservation news and bring intimate views of ocean wildlife into your daily routine.

MAKE INFORMED FOOD CHOICES

Conservation is tricky, and we are not always in a position to make it a priority. If you are fortunate enough to have the resources and time to examine and change your habits, food is a great place to start. If you eat seafood, find out where it's coming from and try to choose more sustainable options. Ask questions at the stores and restaurants where you buy your food. The Monterey Bay Aquarium publishes a free seafood watch list that can help you identify the most ocean-friendly seafood—and what to avoid—anywhere in the United States.

SPEAK OUT

Call or write your representatives to let your government know that conservation is important to you. Whether you live on a coastline near a protected marine area or far inland from any major body of water, your life and your community have an impact on the world's oceans. Express your support for protection of wild spaces on land and in the sea.

DONATE

If you are able, consider making a financial contribution to conservation projects that work with endangered cetacean species, such as the World Wildlife Fund or Mission Blue. Or contribute to your local schools and science programs so more kids have the opportunity to learn about science, conservation, and nature—and grow up to be a more responsible generation.

DON'T GIVE UP

It's easy to become jaded or feel defeated when it comes to environmentalism, but don't give up the fight. At this critical point in the history of humans and nature, it's our time to do what we can to help those who can't help themselves.

SOURCES

BOOKS

Alexander, Becky, ed. *Smithsonian Natural History: The Ultimate Visual Guide to Everything on Earth*. New York: DK Publishing, 2010.

The Animal Book: A Visual Encyclopedia of Life on Earth. New York: DK Children, 2013.

Berta, Annalisa. *Whales, Dolphins, & Porpoises: A Natural History and Species Guide*. Chicago: University of Chicago Press, 2015.

Burnett, D. Graham. *The Sounding of the Whale: Science and Cetaceans in the Twentieth Century*. Chicago: The University of Chicago Press, 2012.

Carwardine, Mark, and Martin Camm. *Whales, Dolphins and Porpoises*. New York: DK Publishing, 2002.

Carwardine, Mark, R. Ewan Fordyce, Peter Gill, and Erich Hoyt. *Whales, Dolphins, & Porpoises*. San Francisco: Fog City Press, 1998.

Kolbert, Elizabeth. *The Sixth Extinction: An Unnatural History*. London: Bloomsbury, 2015.

Stewart, Brent S., Phillip J. Clapham, and James A. Powell. *National Audubon Society Field Guide to Marine Mammals of the World*. New York: A.A. Knopf, 2002.

DOCUMENTARIES

"Beach Babies." *Baby Animals in the Wild*. National Geographic. 2015.

"Blue Whale." *Last Chance to See*. BBC Two. 2009.

"Cape." *Africa*. BBC Natural History Unit. 2013.

David Attenborough's Natural Curiosities. BBC Worldwide. 2013.

Dolphins: Spy in the Pod. BBC One. 2014.

Humpback Whales. Directed by Greg MacGillivray. 2015.

Jane & Payne. Netflix. 2016.

Nature's Great Events. BBC One. 2009.

"Ocean Giants." *Ocean Giants*. BBC One. 2011.

WEBSITES

American Museum of Natural History. www.amnh.org.

California Academy of Sciences. www.calacademy.org.

The Field Museum. www.fieldmuseum.org.

National Geographic. www.nationalgeographic.com.

ACKNOWLEDGMENTS

Thank you to my editor, Kaitlin Ketchum, and designer, Betsy Stromberg, for their expert guidance and support in creating this book. I couldn't do this without them. Thank you also to Ten Speed Press production manager Jane Chinn, publicist Natalie Mulford, design fellow Christine Innes, and publisher Aaron Wehner. Many thanks to Maureen Flannery at the California Academy of Sciences. And thanks as always to my best bud Nick, my parents Jeff and Julie Oseid, and to Danny and Olivia.

ABOUT THE AUTHOR

KELSEY OSEID is an author and illustrator based in the Midwest. Her work is a celebration of science, nature, and the ways humans relate to the natural world. She lives in Minneapolis with her husband, Nick Wojciak, two cats, and two chickens.

INDEX

Library of Congress Cataloging-in-Publication Data
Names: Oseid, Kelsey, author.
Title: Whales : an illustrated celebration / Kelsey Oseid.
Description: First edition. | California : Ten Speed Press,
    [2018] | Includes bibliographical references and index.
Identifiers: LCCN 2018001020
Subjects: LCSH: Whales.
Classification: LCC QL737.C4 O84 2018 |
    DDC 599.5—dc23
LC record available at https://lccn.loc.gov/2018001020

Hardcover ISBN: 978-0-399-58183-0
eBook ISBN: 978-0-399-58184-7

Printed in China

Design by Betsy Stromberg and Christine Innes

10 9 8 7 6 5 4

First Edition